Collins

Edexcel GCSE 9-1

Maths Higher

Practice Papers

Phil Duxbury and Keith Gordon

Contents

SET A

Paper 1 (non-calculator) 3

Paper 2 (calculator) 19

Paper 3 (calculator) 35

SET B

Paper 1 (non-calculator) 51

Paper 2 (calculator) 67

Paper 3 (calculator) 83

ANSWERS 99

Acknowledgements

The authors and publisher are grateful to the copyright holders for permission to use quoted materials and images.

All images are © HarperCollins*Publishers* and Shutterstock.com

Every effort has been made to trace copyright holders and obtain their permission for the use of copyright material. The authors and publisher will gladly receive information enabling them to rectify any error or omission in subsequent editions. All facts are correct at time of going to press.

Published by Collins
An imprint of HarperCollins*Publishers*
1 London Bridge Street
London SE1 9GF

HarperCollins*Publishers*
1st Floor, Watermarque Building,
Ringsend Road, Dublin 4, Ireland

© HarperCollins*Publishers* Limited 2021

ISBN 9780008321499

First published 2019

This edition published 2021

10 9 8 7 6

All rights reserved. No part of this publication may be reproduced, stored in a retrieval system, or transmitted, in any form or by any means, electronic, mechanical, photocopying, recording or otherwise, without the prior permission of Collins.

British Library Cataloguing in Publication Data.

A CIP record of this book is available from the British Library.

Commissioning Editor: Kerry Ferguson
Project Management: Chantal Addy and Richard Toms
Authors: Phil Duxbury and Keith Gordon
Cover Design: Sarah Duxbury and Kevin Robbins
Inside Concept Design: Ian Wrigley
Text Design and Layout: QBS Learning
Production: Karen Nulty
Printed and bound in the UK using 100% Renewable Electricity at CPI Group (UK)

©HarperCollins*Publishers* 2019

Collins

Edexcel
GCSE
Mathematics

H

SET A – Paper 1 Higher Tier (Non-Calculator)

Author: Phil Duxbury

Time allowed: 1 hour 30 minutes

You must have:

- Ruler graduated in centimetres and millimetres, protractor, pair of compasses, pen, HB pencil, eraser.

You may not use a calculator

Instructions

- Use **black** ink or black ball-point pen.
- Answer **all** questions.
- Answer the questions in the spaces provided – *there may be more space than you need.*
- **Calculators may not be used.**
- Diagrams are NOT accurately drawn, unless otherwise indicated.
- You must **show all your working out**.

Information

- The total mark for this paper is 80.
- The marks for **each** question are shown in brackets
  – *use this as a guide as to how much time to spend on each question.*
- Read each question carefully before you start to answer it.
- Keep an eye on the time.
- Try to answer every question.
- Check your answers if you have time at the end.

Name: ....

Answer ALL questions.

Write your answers in the spaces provided.

You must write down all the stages of your working.

1   Find the lowest common multiple of 6, 15 and 40.

(Total for Question 1 is 3 marks)

2   Solve the equation $\dfrac{x-1}{6} = \dfrac{10-x}{3}$

(Total for Question 2 is 3 marks)

3   The plan, front elevation and side elevation of a solid prism are shown below.

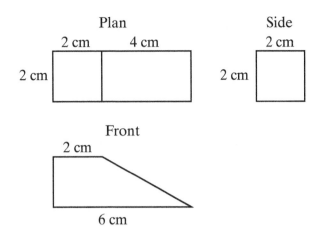

Plan
2 cm   4 cm
2 cm

Side
2 cm
2 cm

Front
2 cm
6 cm

(a) Draw a sketch of the solid prism in 3 dimensions.

(1)

(b) Determine the volume of the prism.

(2)

(Total for Question 3 is 3 marks)

4   Matt wishes to travel from London to Aberdeen, calling in on his friends in Manchester and Glasgow.

From London to Manchester he can either fly, take the train or take a coach.

From Manchester to Glasgow he can either fly, take the train or take a coach.

From Glasgow to Aberdeen, he can either fly or take the train.

In how many different ways can he travel from London to Aberdeen?

(Total for Question 4 is 2 marks)

5   A sequence is generated by the term to term rule 'subtract 5', with the initial term being 100.

(a) Write down the first five terms in the sequence.

(1)

(b) Find a formula for the nth term of the sequence.

(2)

(Total for Question 5 is 3 marks)

6 Write the following numbers in standard form.

(a) 33 000

(1)

(b) 0.0082

(1)

(c) 0.002×10^{-4}

(1)

(Total for Question 6 is 3 marks)

7 The following Venn diagram shows the distribution of 30 random students, all of whom are studying physics, chemistry or biology at GCSE level.

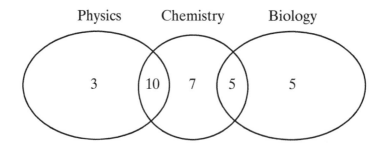

Physics Chemistry Biology

 3 10 7 5 5

(a) Find the probability that a student selected at random studies biology.

                                (1)

(b) Find the probability that a student selected at random studies physics, given that they study chemistry.

                                (1)

(c) Find the probability that a student selected at random studies chemistry, given that they do not study biology.

                                (1)

(Total for Question 7 is 3 marks)

8   Given $p = \dfrac{3-q}{3+q}$, rearrange the formula to make q the subject.

(Total for Question 8 is 3 marks)

9   Expand and simplify the expression $(2x - 1)^3$

(Total for Question 9 is 4 marks)

10   The shape P is enlarged by a scale factor of $-\dfrac{1}{2}$ from the point $(-1,0)$.

Draw the new shape on the grid provided.

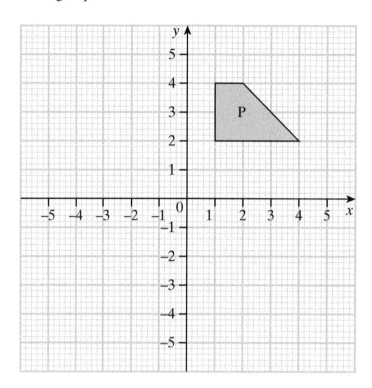

(Total for Question 10 is 2 marks)

---

11   Find the exact values of the following.

(a) $64^{\frac{2}{3}}$

(1)

(b) $\left(\dfrac{16}{25}\right)^{-\frac{3}{2}}$

(2)

(Total for Question 11 is 3 marks)

---

12   The following graph is of the function $y = 2^x$

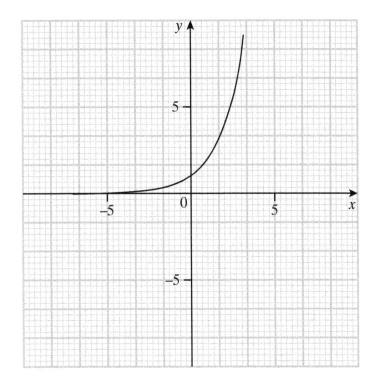

(a) On the same axes, reflect the graph in the line $x = 0$

(1)

(b) State the equation of the new graph.

(1)

(Total for Question 12 is 2 marks)

13   Factorise completely the expression $2x^2 - 32$

(Total for Question 13 is 2 marks)

14   The following diagram shows a circle, centre O.

AB and BC are tangent lines.

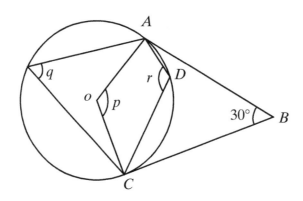

Find the size of the following angles giving your reasons in each case.

$p = $

Reason:

$q = $

Reason:

$r = $

Reason:

(Total for Question 14 is 3 marks)

15 The following box plots illustrate the range of temperatures during one October month for Cyprus and Majorca.

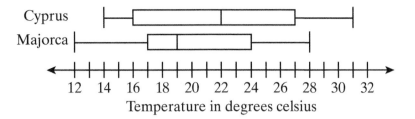

(a) Calculate the interquartile range of temperatures for both Cyprus and Majorca.

(2)

(b) Bill wishes to go on holiday in October, hoping for good weather.

Suggest where he should choose and why.

(2)

(Total for Question 15 is 4 marks)

16 Given the sequence whose general term $u_n = (2\sqrt{3})^n$, find $u_1 + u_2 + u_3 + u_4$, expressing your answer in the form $a + b\sqrt{3}$, where a and b are constants to be determined.

(Total for Question 16 is 4 marks)

17 The ratio of brazil nuts to hazelnuts is $2 : 5$

The ratio of hazelnuts to walnuts is $3 : 7$

(a) Find the ratio of brazil nuts to walnuts.

(3)

(b) If there are 105 walnuts, calculate how many brazil nuts there are.

(1)

(Total for Question 17 is 4 marks)

18 Express the fraction $\dfrac{71}{90}$ as a recurring decimal.

(Total for Question 18 is 3 marks)

19   **(a)**   On the grid below, sketch the graph of $y = \tan x$

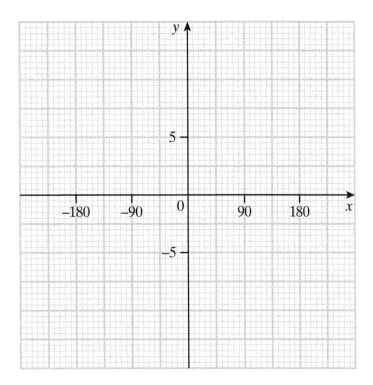

(2)

(b)   Using your graph, solve the equation $\tan x = \sqrt{3}$ for $-180° < x < 180°$

(3)

(Total for Question 19 is 5 marks)

20 Solve the simultaneous equations $y = 2x^2 - 3x + 5$ and $y = 8 - 2x$

21 Prove that $(3n + 1)^2 - (3n - 1)^2$ is a multiple of 6 for all positive integers n.

22 Write each of the following expressions in the form $a + b\sqrt{5}$, where a and b are rational numbers.

(a) $\sqrt{5}(2 - \sqrt{5})^2$

(2)

(b) $\dfrac{5}{5 - 3\sqrt{5}}$

(3)

(Total for Question 22 is 5 marks)

23 **(a)** Sketch the graph of $y = 2x^2 - 3x - 14$ on the grid below, showing clearly where the graph crosses the x and y-axes.

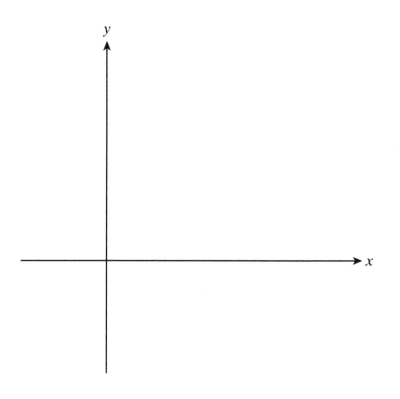

(4)

(b) Solve the inequality $2x^2 - 3x - 14 > 0$, giving your answer in set notation.

(2)

(Total for Question 23 is 6 marks)

TOTAL FOR PAPER IS 80 MARKS

Collins

Edexcel
GCSE
Mathematics

H

SET A – Paper 2 Higher Tier (Calculator)

Author: Phil Duxbury

Time allowed: 1 hour 30 minutes

You must have:

- Ruler graduated in centimetres and millimetres, protractor, pair of compasses, pen, HB pencil, eraser, calculator.

Instructions

- Use **black** ink or black ball-point pen.
- Answer **all** questions.
- Answer the questions in the spaces provided – *there may be more space than you need.*
- **Calculators may be used.**
- Diagrams are NOT accurately drawn, unless otherwise indicated.
- You must **show all your working out**.

Information

- The total mark for this paper is 80.
- The marks for **each** question are shown in brackets
  – *use this as a guide as to how much time to spend on each question.*
- Read each question carefully before you start to answer it.
- Keep an eye on the time.
- Try to answer every question.
- Check your answers if you have time at the end.

Name: _____

Answer ALL questions.

Write your answers in the spaces provided.

You must write down all the stages of your working.

1   The following table shows the heights of giraffes at a zoo.

height (x cm)	frequency
$500 \leqslant x < 510$	2
$510 \leqslant x < 520$	6
$520 \leqslant x < 530$	1
$530 \leqslant x < 540$	4
$540 \leqslant x < 550$	3

(a) State the modal class interval.

....

(1)

(b) Find an estimate for the mean height of the giraffes.

....

(2)

(Total for Question 1 is 3 marks)

2   In the following diagram, $ABCD$ is a cyclic quadrilateral.

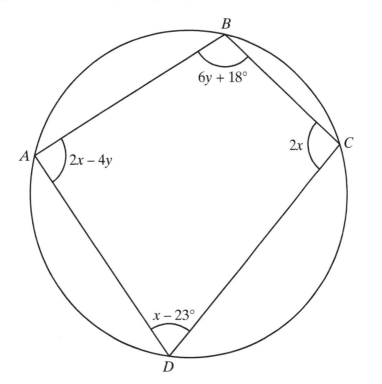

Not drawn accurately

Find the value of x and the value of y.

$x = $ ....

$y = $ ....

(Total for Question 2 is 5 marks)

3   Solve the inequality $-3 < \dfrac{2x+7}{4} < 5$, illustrating your answer on a number line.

(Total for Question 3 is 4 marks)

4   Sadiq invests £1000 in a savings account paying a compound interest rate of 1.25%

For the first year only, there is a bonus 0.75% interest.

Calculate the amount (to the nearest pound) he can expect to have in his account after 5 years.

(Total for Question 4 is 2 marks)

5   A maths test comprises of two papers: paper 1 and paper 2

A student completes paper 1, then tackles paper 2

The probability that a student passes paper 1 is 0.7, and the probability that a student passes paper 2 is 0.8

(a)   Complete the probabilities on the following tree diagram.

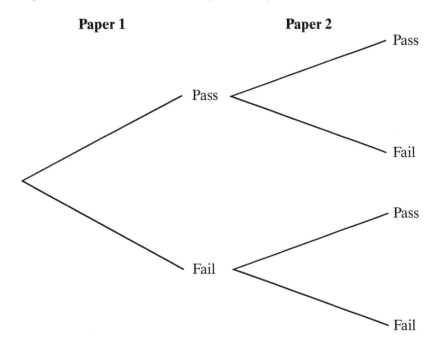

(2)

(b)   Find the probability that the student passes at least one of the papers.

(2)

(Total for Question 5 is 4 marks)

6   Solve the equation $x^2 + x - 90 = 0$

(Total for Question 6 is 3 marks)

7   Given vector $\mathbf{a} = \begin{pmatrix} 3 \\ -2 \end{pmatrix}$ and vector $\mathbf{b} = \begin{pmatrix} -2 \\ -1 \end{pmatrix}$, calculate the vector $2\mathbf{a} - 3\mathbf{b}$.

(Total for Question 7 is 2 marks)

8 In the following triangle, $AB = 20$ cm, $BC = 33$ cm and angle $ABC = 40°$

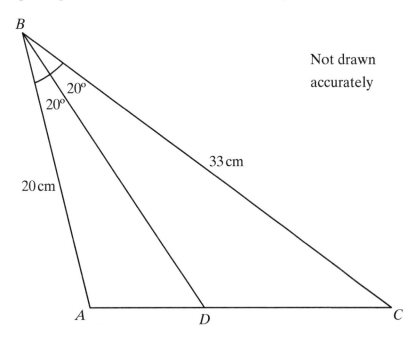

Not drawn
accurately

(a) Find the size of length AC, giving your answer to 3 significant figures.

(3)

The line BD bisects the angle ABC.

(b) Find the ratio of the area of triangle ABD : area of triangle BCD.

(3)

(Total for Question 8 is 6 marks)

9 A quantity y is inversely proportional to the square root of x.

Given $y = 12.5$ when $x = 16$, find the value of y when $x = 0.25$

$y = $ ....

(Total for Question 9 is 4 marks)

---

10 In the diagram below, angle $ABC = 140°$

Using your ruler and compasses only, construct an angle of 35°, making your construction lines clear.

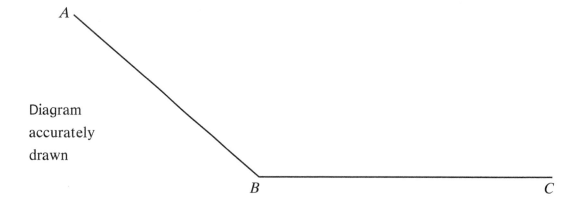

Diagram
accurately
drawn

(Total for Question 10 is 2 marks)

11 **(a)** Find the equation of the following line L, expressing your answer in the form $y = mx + c$

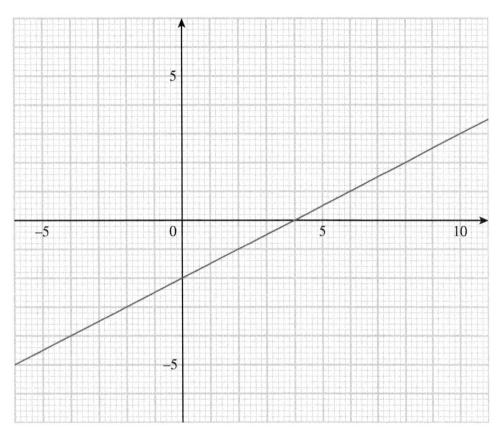

(3)

(b) Find the equation of the line perpendicular to L that intersects at the point $(10, 0)$.

(3)

(Total for Question 11 is 6 marks)

12 Solid A and solid B are mathematically similar.

The ratio of the volume of A to the volume of B is $27 : 125$

Given that the volume of the larger solid is 0.1 m^3, find the surface area of the smaller solid (in cm^2).

(Total for Question 12 is 5 marks)

13 In physics, the resistance (in ohms) of a resistor can be calculated using the formula $R = \dfrac{V^2}{P}$, where V is the potential difference across the resistor (measured in volts) and P is the power dissipated (measured in watts, W).

Given that the potential difference is 12 V (to the nearest volt), and the power is measured at 13.8 W (to 3 significant figures), find lower and upper bounds for the resistance.

Lower bound = ...

Upper bound = ...

(Total for Question 13 is 4 marks)

14 A sample of hedgehogs from a local park were observed and their weight measured.

The following data was tabulated.

Weight (x g)	Frequency	Cumulative frequency
$570 \leqslant x < 590$	5	
$590 \leqslant x < 610$	12	
$610 \leqslant x < 630$	10	
$630 \leqslant x < 650$	8	
$650 \leqslant x < 670$	10	
$670 \leqslant x < 700$	5	

(a) Complete the cumulative frequency column. (1)

(b) On the following grid, draw a cumulative frequency polygon of the data.

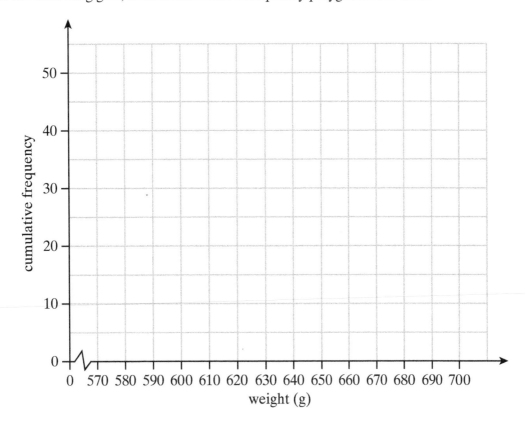

(2)

(c) A hedgehog is deemed to be healthy if its weight is at least 615 g.
Use the cumulative frequency polygon to determine the percentage of healthy hedgehogs in the sample.

....

(3)

(Total for Question 14 is 6 marks)

15 Consider the functions $f(x) = \dfrac{1}{x-2} \, (x \neq 2)$ and $g(x) = x^2 \, (x \geq 0)$

(a) Find an expression for $f^{-1}(x)$

(3)

(b) Find an expression for $gf(x)$

(1)

(c) Solve the equation $fg(x) = gf(x)$

(2)

(Total for Question 15 is 6 marks)

16 *ABCD* is a parallelogram.

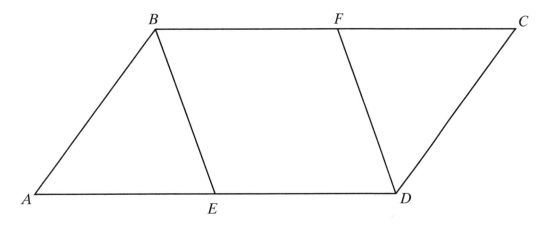

Given $AE = FC$, prove that $BE = FD$.

(Total for Question 16 is 4 marks)

---

17 The following shape shows a solid hemisphere of radius $\sqrt{3}$ cm, affixed to a cone of perpendicular height $2\sqrt{3}$ cm.

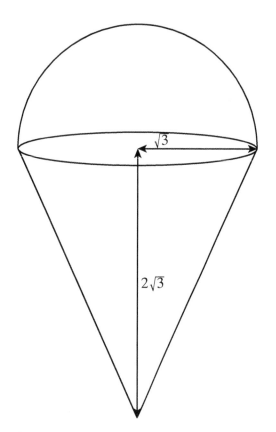

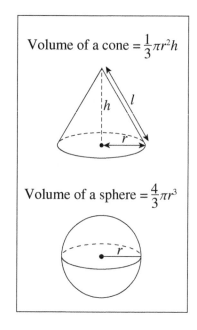

Volume of a cone $= \frac{1}{3}\pi r^2 h$

Volume of a sphere $= \frac{4}{3}\pi r^3$

Find the volume of the composite solid, expressing your answer in the form $a\pi\sqrt{3}$, where a is a constant.

(Total for Question 17 is 5 marks)

18 **(a)** Show that the equation $x^4 + 2x - 7 = 0$ has a solution between $x = 1.3$ and $x = 1.5$

(2)

(b) Starting with $x_0 = 1.4$, use the iteration formula $x_{n+1} = \sqrt[4]{7 - 2x}$ **three** times to find a solution to the equation $x^4 + 2x - 7 = 0$

Give your final answer to 3 decimal places.

(3)

(Total for Question 18 is 5 marks)

19   Find the equation of the line of symmetry of the curve $C: y = 3 - 2x - 4x^2$

Hence or otherwise, find the coordinates of the maximum point on the curve C.

(Total for Question 19 is 4 marks)

TOTAL FOR PAPER IS 80 MARKS

Collins

Edexcel
GCSE
Mathematics

H

SET A – Paper 3 Higher Tier (Calculator)

Author: Phil Duxbury

Time allowed: 1 hour 30 minutes

You must have:

- Ruler graduated in centimetres and millimetres, protractor, pair of compasses, pen, HB pencil, eraser, calculator.

Instructions

- Use **black** ink or black ball-point pen.
- Answer **all** questions.
- Answer the questions in the spaces provided – *there may be more space than you need.*
- **Calculators may be used.**
- Diagrams are NOT accurately drawn, unless otherwise indicated.
- You must **show all your working out**.

Information

- The total mark for this paper is 80.
- The marks for **each** question are shown in brackets
  – *use this as a guide as to how much time to spend on each question.*
- Read each question carefully before you start to answer it.
- Keep an eye on the time.
- Try to answer every question.
- Check your answers if you have time at the end.

Name: ....

1   Find the length of the side marked x in the following right-angled triangle.

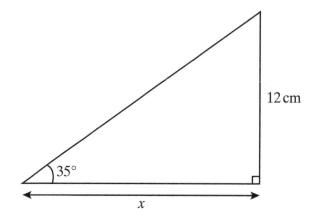

Not drawn accurately

12 cm

35°

x

$x = $ .... cm

(Total for Question 1 is 2 marks)

2   The price of a car is reduced to £21 120 following a reduction of 12%

Find the original price of the car.

£ ....

(Total for Question 2 is 2 marks)

3 Find the length marked x in the following diagram.

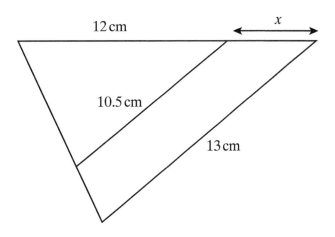

12 cm

x

10.5 cm

13 cm

$x =$ _____ cm

(Total for Question 3 is 3 marks)

4 Change the recurring decimal $0.1\overset{\cdot\cdot}{27}$ into a rational fraction, simplifying your answer as far as possible.

(Total for Question 4 is 4 marks)

5   State which of the following are equations, and which are identities.

Give your reasons.

(a) $(x - 3)^2 = x^2 + 9$

(1)

(b) $\cos(90° - x) = \sin x°$

(1)

(c) $x + 1 = \dfrac{1}{x + 1}$

(1)

(Total for Question 5 is 3 marks)

6   The following shows a velocity–time graph for a cyclist.

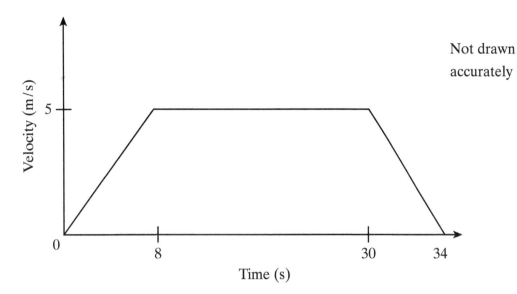

Not drawn accurately

Find the total distance travelled.

(Total for Question 6 is 3 marks)

7 The following sector OAB has a radius of 15 cm and an area of 250 cm^2.

Find the size of the angle θ.

Give your answer to 3 significant figures.

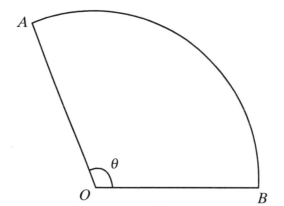

Not drawn accurately

.... °

(Total for Question 7 is 2 marks)

8   GCSE geography students are given a grade for their project work on a scale from 0 to 20.

The following histogram shows the distribution of grades from a cohort of 70 students.

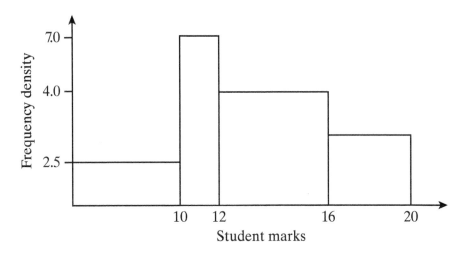

(a) Find the height of the fourth bar.

(3)

(b) Use the histogram to find an estimate for the median student mark.

(2)

(Total for Question 8 is 5 marks)

9 Solve the equation $\dfrac{12}{x-2} - \dfrac{2}{x-3} = 3$, giving your answers to 2 decimal places.

(Total for Question 9 is 5 marks)

10

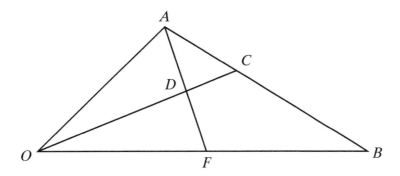

In triangle OAB, $\overrightarrow{OA} = \mathbf{a}$, $\overrightarrow{OB} = \mathbf{b}$. C splits AB in the ratio $1:n$

Find the following vectors in terms of $\mathbf{a}$ and $\mathbf{b}$.

(a) $\overrightarrow{AB}$

(1)

(b) $\overrightarrow{AC}$

(1)

(c) $\overrightarrow{OC}$

(2)

(d) Given D splits AF in the ratio $2:3$ and F bisects OB, find the value of n.

(4)

(Total for Question 10 is 8 marks)

11 The following diagram shows the motion of an object decelerating from a velocity of 6 m/s to rest in 7.75 seconds.

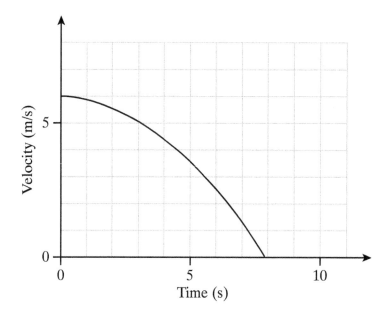

(a) State how you know the deceleration is not constant.

(1)

(b) Find an estimate for the deceleration of the object after 2 seconds.

(4)

(Total for Question 11 is 5 marks)

12 The following diagram shows the graph of $y = x^4 - 5x^2 + 3$

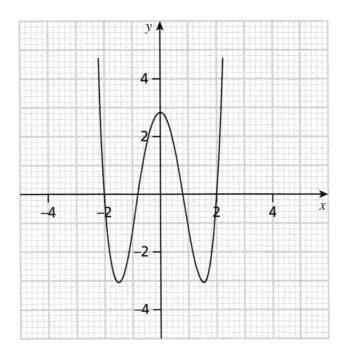

By drawing suitable straight lines, solve the following equations.

(a) $x^4 - 5x^2 - 1 = 0$

(3)

(b) $x^4 - 5x^2 - x + 2 = 0$

(3)

(Total for Question 12 is 6 marks)

13   In a box of 20 coloured counters, 2 are red, 8 are blue and 10 green.

Greg takes three counters from the box, one after the other, without replacing them.

What is the probability that he chooses three counters of different colour?

Give your answer as a simplified fraction.

(Total for Question 13 is 5 marks)

14   **(a)** Express $2x^2 - 6x + 1$ in the form $a(x - b)^2 + c$, where a, b and c are constants.

(4)

(b) Find the range of the function $f(x) = 2x^2 - 6x + 1$

(2)

(Total for Question 14 is 6 marks)

15   The population P (in thousands) of wild geese in a certain area is in decline.

20 months after first analysing the geese, there were estimated to be a population of 8000.
After a further 20 months there were only 1600.

It is thought that the population P (in thousands) of wild geese is related to time (in months)
by the formula $P = A \times 5^{-kt}$, where A and k are both constants.

Use the given information to determine the exact value of A and the exact value of k.

$A =$ ....

$k =$ ....

(Total for Question 15 is 5 marks)

16 A right circular cone has radius r, perpendicular height h and slant height 25 cm.

Given that the total surface area of the cone is 600π cm^2, find the volume of the cone, giving your answer as a multiple of π.

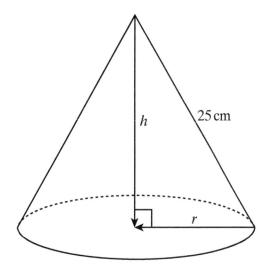

Curved surface area of a cone $= \pi r l$

Volume of a cone $= \frac{1}{3}\pi r^2 h$

(Total for Question 16 is 7 marks)

17 The following diagram shows a circle, centred at O.

A is a point on the circumference of the circle, with coordinates $A(2\sqrt{5}, 6)$.

AB is a tangent to the circle.

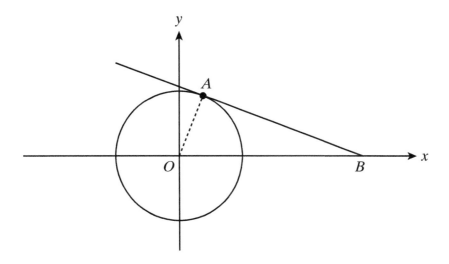

(a) Find the equation of the circle.

(2)

(b) Find the equation of the line AB in the form $y = mx + c$

(4)

(c) Find the area of the triangle OAB in the form $k\sqrt{5}$, where k is a constant to be determined.

(3)

(Total for Question 17 is 9 marks)

TOTAL FOR PAPER IS 80 MARKS

Collins

Edexcel

GCSE

Mathematics

H

SET B – Paper 1 Higher Tier (Non-Calculator)

Author: Keith Gordon

Time allowed: 1 hour 30 minutes

You must have:

- Ruler graduated in centimetres and millimetres, protractor, pair of compasses, pen, HB pencil, eraser.

You may not use a calculator

Instructions

- Use **black** ink or black ball-point pen.
- Answer **all** questions.
- Answer the questions in the spaces provided – *there may be more space than you need.*
- **Calculators may not be used.**
- Diagrams are NOT accurately drawn, unless otherwise indicated.
- You must **show all your working out**.

Information

- The total mark for this paper is 80.
- The marks for **each** question are shown in brackets
  – *use this as a guide as to how much time to spend on each question.*
- Read each question carefully before you start to answer it.
- Keep an eye on the time.
- Try to answer every question.
- Check your answers if you have time at the end.

Name: _____

1    $f(x) = x - 3$

Write down an expression for $f^{-1}(x)$

(Total for Question 1 is 1 mark)

2   Write down the roots of the equation $(x - 2)(x + 3) = 0$

(Total for Question 2 is 1 mark)

3   Here is a right-angled triangle ABC.

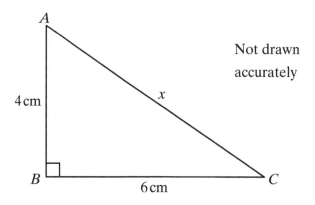

Not drawn accurately

Work out the **exact** value of the length x.

$x =$ _____ cm

(Total for Question 3 is 2 marks)

4   Solve $3(x - 2) + 4 = \dfrac{x}{2}$

(Total for Question 4 is 3 marks)

5   Work out the surface area of the cuboid shown.

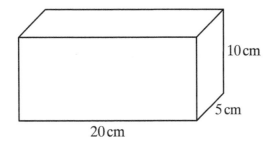

10 cm

5 cm

20 cm

.... cm^2

(Total for Question 5 is 3 marks)

6   Expand and simplify $4(x + 1) - 2(3x - 4)$

(Total for Question 6 is 3 marks)

7   Part of a regular polygon is shown.

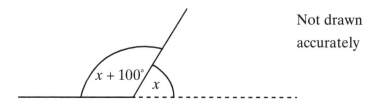

Not drawn
accurately

How many sides does the polygon have?

(Total for Question 7 is 3 marks)

8   **(a)**   Write 2.3×10^5 as an ordinary number.

....

                                                            (1)

(b)   Write 0.0005 in standard form.

....

                                                            (1)

(c)   Work out $2 \times 10^4 \times 8 \times 10^3$

Give your answer in standard form.

....

                                                            (2)

(Total for Question 8 is 4 marks)

9   The graph of $y = 2x^2 - 3x - 5$ is shown.

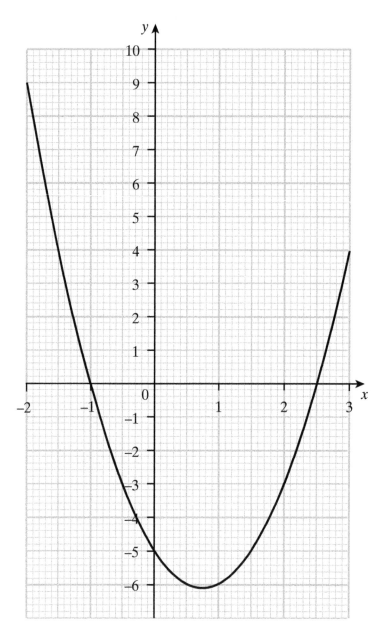

(a) Write down the values of x when $y = 4$.

....

(2)

(b) Write down the coordinates of the minimum point.

....

(1)

(Total for Question 9 is 3 marks)

10 Here is a square.

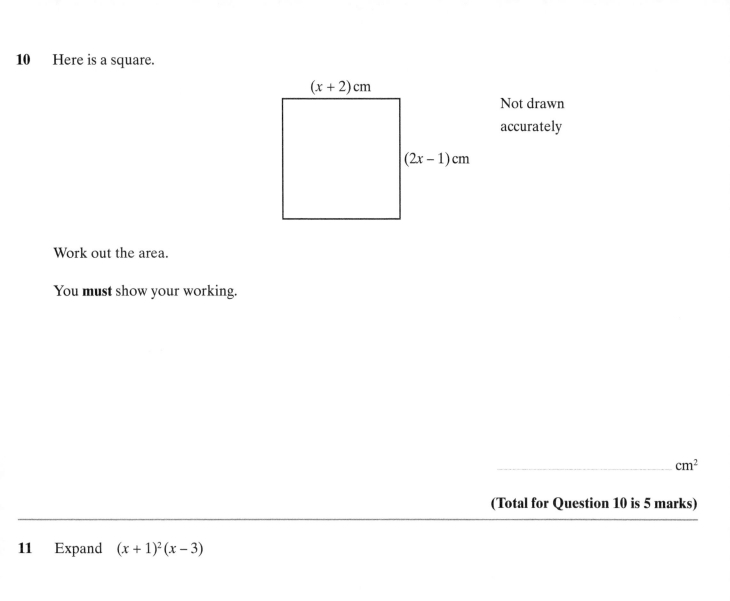

$(x + 2)$ cm

Not drawn
accurately

$(2x - 1)$ cm

Work out the area.

You **must** show your working.

.... cm²

(Total for Question 10 is 5 marks)

11 Expand $(x + 1)^2 (x - 3)$

....

(Total for Question 11 is 3 marks)

12 A cylinder has a base diameter that is $\frac{1}{3}$ of the height.

The volume of the cylinder is 48π

Work out the **radius** of the base.

(Total for Question 12 is 3 marks)

13

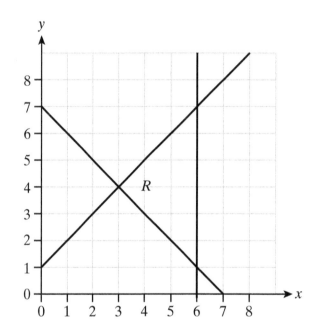

Write down the three inequalities that define the region R.

(Total for Question 13 is 3 marks)

14 Expand and simplify $(3+\sqrt{2})(9-\sqrt{8})$

Give your answer in the form $a+b\sqrt{2}$, where a and b are integers.

(Total for Question 14 is 3 marks)

15 Draw a histogram for the data below.

Height, h cm	Frequency
$5 \leqslant h < 10$	15
$10 \leqslant h < 20$	35
$20 \leqslant h < 35$	30
$35 \leqslant h < 45$	15
$45 \leqslant h < 50$	5

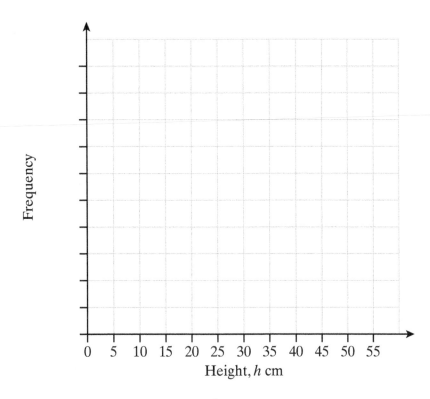

(Total for Question 15 is 3 marks)

16 **(a)** *O* is the centre of the circle.

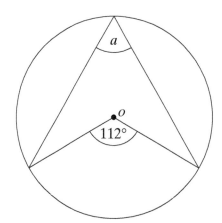

Not drawn
accurately

Write down the size of angle *a* in degrees.

....°

(1)

(b) *O* is the centre of the circle.

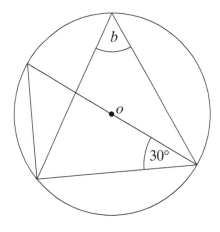

Not drawn
accurately

Write down the size of angle *b* in degrees.

....°

(1)

(c) *ABC* are points on the circumference of a circle, centre *O*.

SAT is a tangent.

BC is a diameter.

Angle *BAT* = 32°

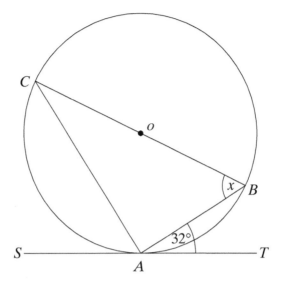

Not drawn
accurately

Work out the size of angle *CBA*, marked *x* on the diagram.

You **must** show your working, which may be on the diagram.

x = ... °

(3)

(Total for Question 16 is 5 marks)

17   Work out $64^{\frac{2}{3}}$

(Total for Question 17 is 2 marks)

18 The cumulative frequency diagram shows the ages of people at a wedding.

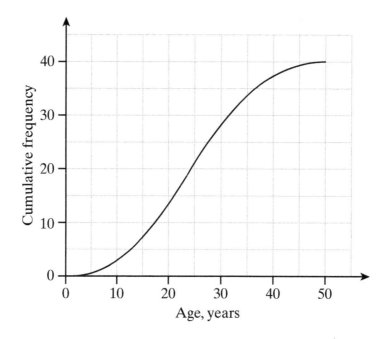

(a) Write down an estimate of the median age.

....

(1)

(b) Work out an estimate of the interquartile range.

....

(2)

(c) The youngest person at the wedding was 5 years old.

Draw a box plot for the data.

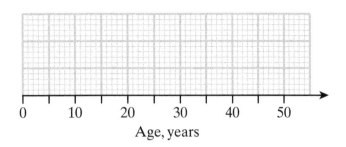

Age, years

(3)

(Total for Question 18 is 6 marks)

---

19 $OABC$ is a trapezium.

$\overrightarrow{OA} = \mathbf{a}$

$\overrightarrow{AB} = \dfrac{3}{2}\mathbf{b}$

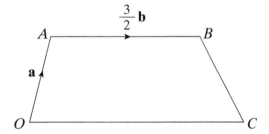

Not drawn
accurately

(a) Write down the vector $\overrightarrow{OB}$ in terms of $\mathbf{a}$ and $\mathbf{b}$.

(1)

(b) $\overrightarrow{BC} = -\mathbf{a} + \dfrac{1}{2}\mathbf{b}$

Work out the vector $\overrightarrow{OC}$.

(2)

(Total for Question 19 is 3 marks)

20   Write the recurring decimal 3.733333…. as a mixed number.

(Total for Question 20 is 3 marks)

21   The area of a right-angled isosceles triangle is 9 cm²

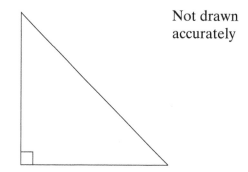

Not drawn
accurately

Work out the perimeter of the triangle.

Give your answer in the form $a + b\sqrt{c}$, where a, b and c are integers.

(Total for Question 21 is 5 marks)

22 A bag contains 10 counters.

7 of them are red, 3 of them are blue.

Two counters are taken from the bag.

Work out the probability that they are different colours.

(Total for Question 22 is 4 marks)

23 Simplify fully $\dfrac{4x^2-4x-15}{2x+8}\times\dfrac{2x^2+5x-12}{4x^2-9}$

(Total for Question 23 is 4 marks)

24 $A(3, 10)$ and B(7, 8) are two points.

Work out the equation of the line that is

perpendicular to AB

passes through the midpoint of AB.

(Total for Question 24 is 5 marks)

TOTAL FOR PAPER IS 80 MARKS

Collins

Edexcel

GCSE

Mathematics

H

SET B – Paper 2 Higher Tier (Calculator)

Author: Keith Gordon

Time allowed: 1 hour 30 minutes

You must have:

- Ruler graduated in centimetres and millimetres, protractor, pair of compasses, pen, HB pencil, eraser, calculator.

Instructions

- Use **black** ink or black ball-point pen.
- Answer **all** questions.
- Answer the questions in the spaces provided – *there may be more space than you need.*
- **Calculators may be used.**
- Diagrams are NOT accurately drawn, unless otherwise indicated.
- You must **show all your working out.**

Information

- The total mark for this paper is 80.
- The marks for **each** question are shown in brackets
  – *use this as a guide as to how much time to spend on each question.*
- Read each question carefully before you start to answer it.
- Keep an eye on the time.
- Try to answer every question.
- Check your answers if you have time at the end.

Name: ....

1   The point $A(6, 7)$ is reflected to the point A' in the line $y = x$.

Work out the coordinates of A'.

(Total for Question 1 is 2 marks)

2   Here are four straight lines, two of which are parallel.

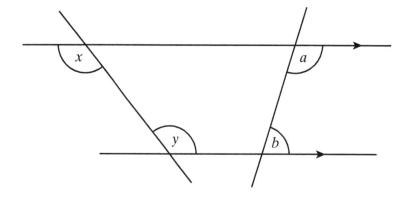

(a) Complete the sentence with the correct word that describes the relationship between angle x and angle y.

Angle x and angle y are _____ angles.   **(1)**

(b) Write down an equation that describes the relationship between angle a and angle b.

(1)

(Total for Question 2 is 2 marks)

3   Translate the triangle by $\begin{pmatrix} -3 \\ -4 \end{pmatrix}$

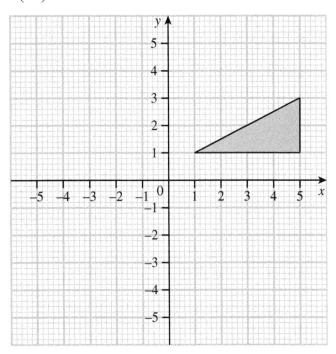

(Total for Question 3 is 2 marks)

4 Work out the length x in the triangle.

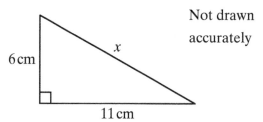

6 cm

Not drawn
accurately

x

11 cm

$x = $ .... cm

(Total for Question 4 is 3 marks)

5 The table shows the heights of some young trees.

Height, h cm	Frequency
$140 \leqslant h < 150$	5
$150 \leqslant h < 160$	9
$160 \leqslant h < 170$	12
$170 \leqslant h < 180$	8
$180 \leqslant h < 190$	6

Work out an estimate of the mean height.

....

(Total for Question 5 is 3 marks)

6 **(a)** As a product of prime factors $20 = 2^2 \times 5$

Work out 28 as a product of prime factors.

....

(2)

(b) Work out the least common multiple of 20 and 28.

....

(2)

(Total for Question 6 is 4 marks)

---

7 Triangles ABC and PQR are similar.

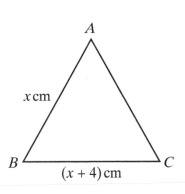 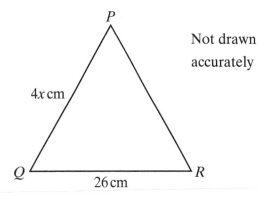

Not drawn accurately

Work out the value of x.

$x = $ ....

(Total for Question 7 is 3 marks)

---

8   A washing machine is reduced by 15% in a sale.

The sale price of the washing machine is £238.

What was the original price of the washing machine?

9   Two numbers are in the ratio 2 : 5

The difference between the numbers is 36.

Work out the values of the two numbers.

10 The area of this semicircle is 201 cm² to 3 significant figures.

Not drawn accurately

Work out the perimeter of the semicircle.

(Total for Question 10 is 3 marks)

11 Using ruler and compasses only, construct an angle of 30° at A.

You must show your construction arcs.

A _____

(Total for Question 11 is 3 marks)

12 **(a)** Expand $5(x-2)(4x+3)$.

(2)

(b) Factorise fully $2x^2 + 8x + 6$.

(2)

(Total for Question 12 is 4 marks)

13 Enlarge the triangle by a scale factor of $-\frac{1}{3}$ about the centre $(5, 8)$.

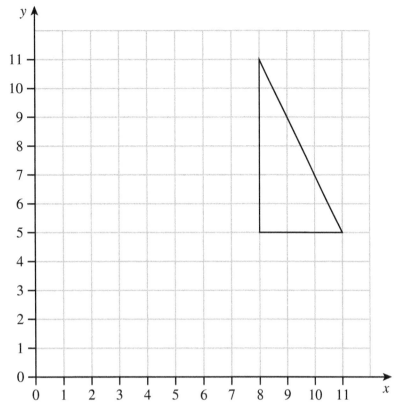

(Total for Question 13 is 3 marks)

---

14 A jar contains 30 red beads and 40 white beads.

The number of red beads is increased by 60%

The number of white beads is increased by $p\%$

The number of red and white beads is now equal.

Work out the value of p.

(Total for Question 14 is 3 marks)

---

15 **(a)** Write $x^2 + 6x - 9$ in the form $(x + a)^2 - b$, where a and b are integers.

(3)

(b) Hence, or otherwise, solve $x^2 + 6x - 9 = 0$

Give answers in the form $p \pm \sqrt{q}$, where p and q are integers.

(2)

(Total for Question 15 is 5 marks)

16 Write the equation $\dfrac{2}{x+1} - \dfrac{3}{4x-1} = 1$

in the form $ax^2 + bx + c = 0$ where a, b and c are integers.

(Total for Question 16 is 4 marks)

17 y is directly proportional to the square of x.

When $y = 20, x = 2$

(a) Work out the value of y when $x = 10$

(3)

(b) Work out the value of x when $y = 5$

(2)

(Total for Question 17 is 5 marks)

18   146 students in year 7 were asked if they had a cat, a dog or both.

The Venn diagram shows the results.

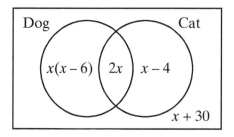

A student is picked at random.

Work out the probability that the student only has a cat.

(Total for Question 18 is 5 marks)

19   **(a)** Work out angle x in this triangle.

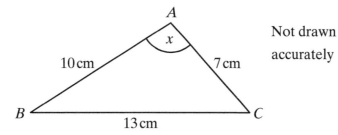

Not drawn
accurately

$x =$ .... °

(3)

(b) Work out the area of this triangle.

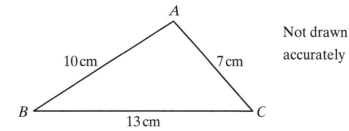

Not drawn
accurately

.... cm^2

(2)

(Total for Question 19 is 5 marks)

20 Show that $\dfrac{6}{3-\sqrt{3}}$ can be simplified to $(3+\sqrt{3})$

You **must** show **all** the steps of your working.

(Total for Question 20 is 3 marks)

---

21 The formula connecting the sine of angle x, the opposite side (o) and the hypotenuse (h) is

$$\sin x = \frac{o}{h}$$

$h = 12$ to 2 significant figures

$o = 8.3$ to 2 significant figures

Work out the upper and lower bounds for the angle x.

Give your angles to 1 decimal place.

You **must** show your working.

Upper bound ....

Lower bound ....

(Total for Question 21 is 5 marks)

22   The speed–time graph for a journey is shown.

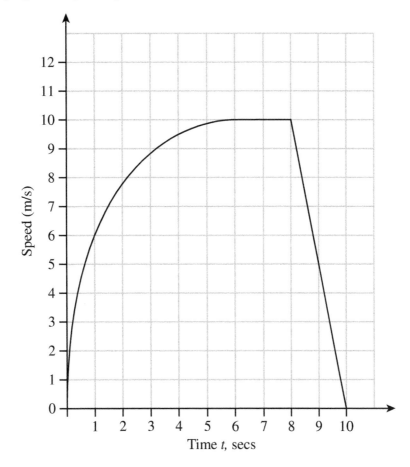

Time t, secs

(a) Estimate the acceleration at 3 seconds.

(3)

(b) Estimate the average speed for the journey.

(4)

(Total for Question 22 is 7 marks)

TOTAL FOR PAPER IS 80 MARKS

BLANK PAGE

Collins

Edexcel
GCSE
Mathematics

H

SET B – Paper 3 Higher Tier (Calculator)

Author: Keith Gordon

Time allowed: 1 hour 30 minutes

You must have:

- Ruler graduated in centimetres and millimetres, protractor, pair of compasses, pen, HB pencil, eraser, calculator.

Instructions

- Use **black** ink or black ball-point pen.
- Answer **all** questions.
- Answer the questions in the spaces provided – *there may be more space than you need.*
- **Calculators may be used.**
- Diagrams are NOT accurately drawn, unless otherwise indicated.
- You must **show all your working out**.

Information

- The total mark for this paper is 80.
- The marks for **each** question are shown in brackets
  – *use this as a guide as to how much time to spend on each question.*
- Read each question carefully before you start to answer it.
- Keep an eye on the time.
- Try to answer every question.
- Check your answers if you have time at the end.

Name: ....

Answer ALL questions.

Write your answers in the spaces provided.

You must write down all the stages of your working.

1   From this list write down the cube number.

     81   225   729   1024

(Total for Question 1 is 1 mark)

2   From this list write down the power of 5.

     55   100   125   225

(Total for Question 2 is 1 mark)

3   **(a)** Simplify $x^3 \times x^6$

(1)

(b) Simplify $x^{12} \div x^2$

(1)

(Total for Question 3 is 2 marks)

4   Here are two column vectors.

$$\mathbf{a} = \begin{pmatrix} 2 \\ 3 \end{pmatrix} \quad \mathbf{b} = \begin{pmatrix} 6 \\ -2 \end{pmatrix}$$

Work out $2\mathbf{a} + \mathbf{b}$

(Total for Question 4 is 2 marks)

5   Two inequalities are shown.

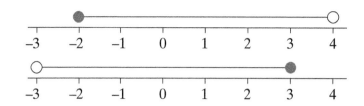

Write down the integers that are in **both** inequalities.

(Total for Question 5 is 2 marks)

6   The diagram shows a right-angled triangle.

One of the other angles is 60°

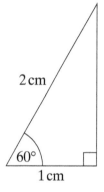

Not drawn
accurately

Calculate the **exact** value of sin 60°

(Total for Question 6 is 2 marks)

7 Enlarge the shape by a scale factor of $\frac{1}{3}$

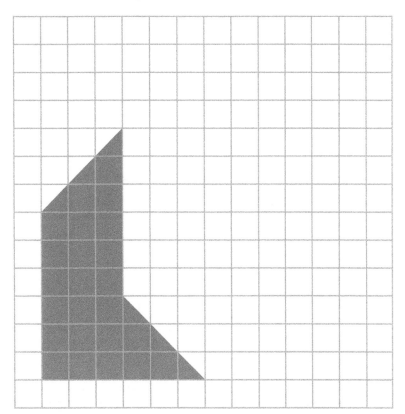

(Total for Question 7 is 2 marks)

---

8 A large candle exerts a pressure of 2 Pa on its base.

As the candle burns the pressure decreases.

After 2 hours the pressure is 0.5 Pa.

Work out the rate of change of pressure.

Give your answer in Pa/hour.

(Total for Question 8 is 2 marks)

9   A bag contains 10 balls.

4 of the balls are red and 6 are blue.

A ball is taken at random from the bag.

The ball is replaced and another ball is taken at random from the bag.

(a) Complete the tree diagram.

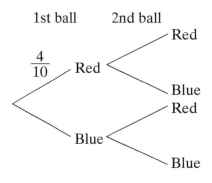

(1)

(b) Use the tree diagram, or otherwise, to work out the probability that both balls were the same colour.

(3)

(Total for Question 9 is 4 marks)

10   Solve the simultaneous equations.

$3x + 2y = 2$

$x + 4y = 9$

(Total for Question 10 is 3 marks)

11 **(a)** Factorise $x^2 - 25$

(1)

(b) Show that $(x + 2)^2 - (x + 1)^2 \equiv 2x + 3$

(3)

(Total for Question 11 is 4 marks)

12 **(a)** Show that the length x in the triangle below is 6.36 cm to 2 decimal places.

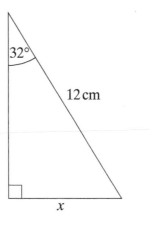

32°

12 cm

Not drawn
accurately

x

(1)

(b) A cone has a half vertical angle of 32° and a slant height *l* of 12 cm.

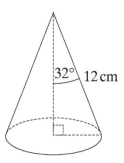

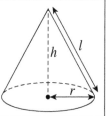

Curved surface area of a cone = πrl

Work out the curved surface area of the cone.

_____ cm^2

(2)

(Total for Question 12 is 3 marks)

---

13   A seal colony has 6000 seals.

It is declining at a rate of 8% per year.

How long will it be before the colony is half its original size?

_____

(Total for Question 13 is 3 marks)

---

14 Match each graph to the equations.

Graph A   Graph B   Graph C

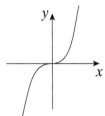 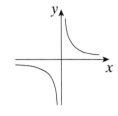 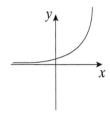

$y = \tan x$ matches graph ....

$y = 2^x$ matches graph ....

$y = \dfrac{1}{x}$ matches graph ....

(Total for Question 14 is 2 marks)

15 Simplify $(2x^2 y^3)^2$

(Total for Question 15 is 2 marks)

16   Here are the equations of four lines.

Line A: $y = 3x + 3$ Line B: $y = \dfrac{1}{4}x - 3$

Line C: $y = \dfrac{1}{3}x + 3$ Line D: $y = -4x - 4$

(a) Which two lines are perpendicular?

....

(1)

(b) Which two lines intersect on the **x-axis**?

....

(1)

(Total for Question 16 is 2 marks)

---

17   **(a)** Write down the next two terms of this quadratic sequence.

3   5   8   12   17   23   ...   ...

....

(2)

(b) Work out the nth term of the quadratic sequence

6   10   16   24   34   46   ...

....

(4)

(Total for Question 17 is 6 marks)

18 The triangle A, shown, is reflected in $y = 6$

Call this triangle B.

Triangle B is then reflected in $x = 5$

Call this triangle C.

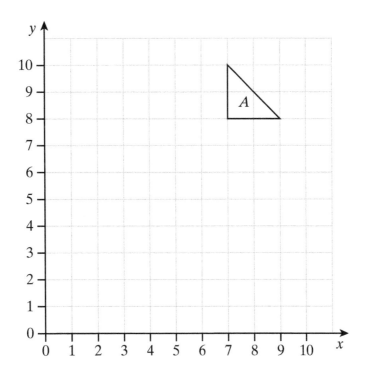

Describe the **single** transformation that will map triangle C to triangle A.

....

....

(Total for Question 18 is 4 marks)

19 Work out the length x in the triangle.

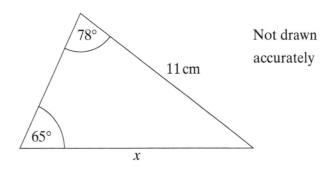

78°

11 cm

Not drawn
accurately

65°

x

$x = $ .... cm

(Total for Question 19 is 3 marks)

20 These two bottles are similar in shape.

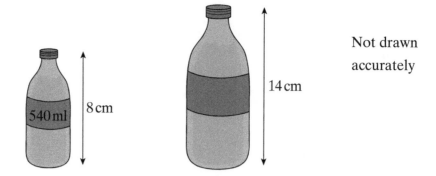

Not drawn accurately

Work out the volume of the large bottle.

Give your answer to 3 significant figures.

(Total for Question 20 is 3 marks)

21 A pyramid has a rectangular base *ABCD*.

The vertex is directly over the midpoint, *X*, of the base.

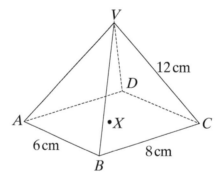

Calculate the angle between the side *VC* and the base *ABCD*.

(Total for Question 21 is 5 marks)

©HarperCollins*Publishers* 2019

22   **(a)** Rearrange the equation $b^3 - 2a + 3 = 0$ to make b the subject.

(1)

(b) One solution of the equation $x^3 - 2x + 3 = 0$ can be found with the iterative formula

$$x_{n+1} = \sqrt[3]{2x_n - 3}$$

Starting with $x_0 = 1$, write down the value of x_1

(1)

(c) Continue the iteration to find the solution.

Give your answer to 2 decimal places.

(2)

(Total for Question 22 is 4 marks)

23 A circle and a line are shown on the centimetre grid.

The line intersects the circle at A.

The circle intersects the x-axis at B.

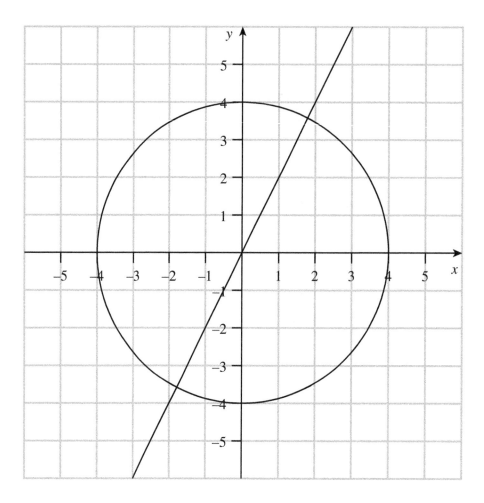

(a) Write down the equation of the circle.

....

(1)

(b) Work out the length of the minor arc AB.

....

(3)

(Total for Question 23 is 4 marks)

24 There are x beads in a jar.

The probability of taking a red bead from the jar at random is $\dfrac{4}{9}$

7 more red beads are added to the jar.

The probability of taking a red bead from the jar at random is now $\dfrac{1}{2}$

Use algebra to work out the value of x.

(Total for Question 24 is 5 marks)

25 Two functions are $f(x) = 3x - 1$ and $g(x) = x^2 + 2$

(a) Work out $f^{-1}(x)$

(2)

(b) Work out $fg(x)$

(2)

(Total for Question 25 is 4 marks)

26 Solve the simultaneous equations

$$y = x + 3$$

$$x^2 + y^2 = x + 12$$

(Total for Question 26 is 5 marks)

TOTAL FOR PAPER IS 80 MARKS

Answers

Key to abbreviations used within the answers

M   method mark (e.g. M1 means 1 mark for method)

A   accuracy mark (e.g. A1 means 1 mark for accuracy)

B   independent marks that do not require method to be shown (e.g. B2 means 2 independent marks)

dep   dependent on previous mark

ft   follow through

oe   or equivalent

Set A – Paper 1

Question	Answer	Mark
1	$6 = 2 \times 3$ $15 = 3 \times 5$ $40 = 2 \times 2 \times 2 \times 5$ LCM is $2 \times 2 \times 2 \times 3 \times 5 = 120$	M1 M1 A1
2	$3(x-1) = 6(10-x)$ $3x - 3 = 60 - 6x$ $9x = 63$ $x = 7$	M1 A1 A1
3 (a)		B1
(b)	$\frac{1}{2} \times (6+2) \times 2 \times 2 = 16 \text{ cm}^3$	M1 A1
4	$3 \times 3 \times 2 = 18$	M1 A1
5 (a)	100, 95, 90, 85, 80	B1
(b)	Common difference of -5 Sequence is $-5n + c \quad c = 105$ Formula is $-5n + 105$ **or** $105 - 5n$	M1 A1
6 (a)	3.3×10^4	B1
(b)	8.2×10^{-3}	B1
(c)	2×10^{-7}	B1
7 (a)	$\frac{10}{30} = \frac{1}{3}$	B1
(b)	$\frac{10}{22} = \frac{5}{11}$	B1
(c)	$\frac{17}{20}$	B1
8	$p(3+q) = 3 - q$ $3p + pq = 3 - q$ $pq + q = 3 - 3p$ $q(p+1) = 3 - 3p$ $q = \frac{3-3p}{1+p}$ or $\frac{3(1-p)}{1+p}$	M1 M1 A1
9	$(2x-1)^3 = (2x-1)(2x-1)^2$ $= (2x-1)(4x^2 - 4x + 1)$ $= 2x(4x^2 - 4x + 1) - 1(4x^2 - 4x + 1)$ $= 8x^3 - 8x^2 + 2x - 4x^2 + 4x - 1$ $= 8x^3 - 12x^2 + 6x - 1$	M1 A1 M1 A1

Question	Answer	Mark
10	 Correct sized shape Correctly positioned	B1 B1
11 (a)	16	B1
(b)	$= \left(\frac{25}{16}\right)^{\frac{3}{2}} = \left(\frac{5}{4}\right)^3$ $= \frac{125}{64}$	M1 A1
12 (a)		B1
(b)	$y = 2^{-x}$	B1
13	$2(x^2 - 16)$ $= 2(x+4)(x-4)$	B1 B1
14	$p = 150°$ radius meets tangent at 90°; OABC quadr so $360 - (90 + 90)$ $q = 75°$ angle subtended at centre is twice the angle subtended at the circumference $r = 105°$ opposite angles in cyclic quad sum to 180°	B1 B1 B1
15 (a)	Cyprus: $27 - 16 = 11$ degrees Majorca: $24 - 17 = 7$ degrees	B1 B1
(b)	Choose Cyprus because the median temperature is highest (comparing median temperatures $(22 > 19)$) OR choose Majorca because the temperatures are more consistent (comparing IQ ranges $(7 < 11)$)	B1 B1 B1 B1

Question	Answer	Mark
16	$u_1 = 2\sqrt{3}$ $u_2 = 12$ $u_3 = 24\sqrt{3}$ $u_4 = 144$ $u_1 + u_2 + u_3 + u_4 = 2\sqrt{3} + 12 + 24\sqrt{3} + 144$ $= 156 + 26\sqrt{3}$	B1 B1 M1 A1
17 (a)	$2:5 = 6:15$ $3:7 = 15:35$ Brazil : walnut $= 6:35$	B1 B1 B1
(b)	$105 \times \dfrac{6}{35} = 18$ brazil OR $105 \times \dfrac{3}{7} \times \dfrac{2}{5} = 18$ brazil	B1 B1
18	Use short division method to evaluate $7.1 \div 9$ (or equivalent) $0.78888...$ $0.7\dot{8}$	M1 A1 A1
19 (a)	Shape Asymptotes at $\pm 90°$ 	B1 B1
(b)	Draw line through approximately $y = 1.73$ $x = 60°$ $x = -120°$	 M1 A1 A1
20	$2x^2 - 3x + 5 = 8 - 2x$ $2x^2 - x - 3 = 0$ $(2x-3)(x+1) = 0$ $2x - 3 = 0 \Rightarrow x = \dfrac{3}{2}$ and $y = 5$ $x + 1 = 0 \Rightarrow x = -1$ and $y = 10$	B1 M1 A1 A1 A1 A1 A1
21	$(3n+1)^2 - (3n-1)^2$ $= 9n^2 + 6n + 1 - (9n^2 - 6n + 1)$ $= 12n$ $= 6 \times 2n$ so is a multiple of 6	 M1 A1 A1
22 (a)	$\sqrt{5}(2 - \sqrt{5})^2 = \sqrt{5}(9 - 4\sqrt{5})$ $= -20 + 9\sqrt{5}$	B1 B1
(b)	$\dfrac{5}{5 - 3\sqrt{5}} = \left(\dfrac{5}{5 - 3\sqrt{5}}\right)\left(\dfrac{5 + 3\sqrt{5}}{5 + 3\sqrt{5}}\right)$ $= \dfrac{25 + 15\sqrt{5}}{25 - 45} = \dfrac{25 + 15\sqrt{5}}{-20}$ $= -\dfrac{5}{4} - \dfrac{3}{4}\sqrt{5}$	M1 A1 A1
23 (a)	 Correct shape and orientation Intersection points marked at: $(-2, 0) \left(\dfrac{7}{2}, 0\right) (0, -14)$	 B1 A1 A1 A1
(b)	Solution is $(-\infty, -2) \cup \left(\dfrac{7}{2}, \infty\right)$	B1 B1

Set A – Paper 2

Question	Answer	Mark
1 (a)	$510 \leqslant x < 520$ (cm)	B1
(b)	$\sum \dfrac{fx}{f} = \dfrac{\left(\begin{array}{l}505 \times 2 + 515 \times 6 + 525 \\ \times 1 + 535 \times 4 + 545 \times 3\end{array}\right)}{16}$	M1
	$= \dfrac{8400}{16} = 525$ cm	A1
2	$(6y+18)+(x-23)=180$	M1
	$x+6y=185 \qquad (1)$	A1
	$2x+(2x-4y)=180$	
	$x-y=45 \qquad (2)$	A1
	Attempt to solve (1) and (2) simultaneously:	M1
	$x=65^\circ,\ y=20^\circ$	A1
3	$-3 < \dfrac{2x+7}{4} \Rightarrow x > -\dfrac{19}{2}$	B1
	$\dfrac{2x+7}{4} < 5 \Rightarrow x < \dfrac{13}{2}$	B1
	Solution is $-\dfrac{19}{2} < x < \dfrac{13}{2}$	B1
	(number line graph with open circles at $\dfrac{-19}{2}$ and $\dfrac{13}{2}$)	B1
4	$1000 \times 1.02 \times 1.0125^4 = \pounds1072$	M1 A1
5 (a)	Paper 1: 0.7, 0.3	B1
	Paper 2: 0.8, 0.2, 0.8, 0.2	B1
(b)	$1-(0.3 \times 0.2) = 0.94$ (or $0.8 \times 0.7 + 0.8 \times 0.3 + 0.2 \times 0.7$)	M1 A1
6	$(x+10)(x-9)=0$	M1
	$x=-10$ or $x=9$	A1 A1
7	$2\begin{pmatrix}3\\-2\end{pmatrix} - 3\begin{pmatrix}-2\\-1\end{pmatrix} = \begin{pmatrix}6\\-4\end{pmatrix} + \begin{pmatrix}6\\3\end{pmatrix}$	M1
	$= \begin{pmatrix}12\\-1\end{pmatrix}$	A1
8 (a)	$AC^2 = 20^2 + 33^2 - 2 \times 20 \times 33 \times \cos 40^\circ$	M1 A1
	$AC^2 = 477.82\ldots$	
	$AC = 21.9$ cm	A1
(b)	Area $ABD = \dfrac{1}{2} \times 20 \times BD \times \sin 20^\circ$	B1
	Area $BDC = \dfrac{1}{2} \times BD \times 33 \times \sin 20^\circ$	B1
	So ratio area of triangle ABD : area of triangle $BCD = 20:33$	B1

Question	Answer	Mark
9	$y \propto \dfrac{1}{\sqrt{x}}$	
	$y = \dfrac{k}{\sqrt{x}}$	B1
	Substitute $x=16, y=12.5 \Rightarrow k=50$	M1
	$y = \dfrac{50}{\sqrt{x}}$	A1
	Substitute $x=0.25 \Rightarrow y=100$	A1
10	Bisect angle ABC with construction lines	B1
	Bisect the angle just constructed (with construction lines)	B1
11 (a)	$m = \dfrac{2}{4} = \dfrac{1}{2}$	B1
	$c = -2$	B1
	$y = \dfrac{1}{2}x - 2$	B1
(b)	Gradient of new line is -2	B1
	Equation is $y = -2x + c$	M1
	Substituting $x=10, y=0$	
	$0 = -20 + c$	
	$c = 20$	A1
	$y = -2x + 20$	
12	$0.1\text{m}^3 = 100\,000$ cm^3	B1
	Length scale factor is $\sqrt[3]{27} : \sqrt[3]{125} = 3:5$	B1
	Area scale factor is $3^2 : 5^2 = 9:25$	B1
	Required surface area is	
	$100\,000 \times \dfrac{9}{25} = 36\,000$ cm^2	M1 A1
13	Using $11.5 < V < 12.5$	
	And $13.75 < P < 13.85$	
	$R_{\min} = \dfrac{11.5^2}{13.85} = 9.55\,\Omega$	M1 A1
	$R_{\max} = \dfrac{12.5^2}{13.75} = 11.4\,\Omega$	M1 A1

Question	Answer	Mark
14 (a)	5, 17, 27, 35, 45, 50	B1
(b)	Plot (590, 5), (610, 17)… etc. and join consecutive points with straight lines or a curve	B1 B1
(c)	Draw line from 615 on x-axis, up to line, and across (left) to intersect y-axis at approx. 20	M1
	20 hedgehogs underweight implied $50 - 20 = 30$ healthy hedgehogs	A1
	Percentage of healthy hedgehogs is $\left(\dfrac{30}{50}\right) \times 100 = 60\%$	A1
15 (a)	$y = \dfrac{1}{x-2}$ $(x-2)y = 1$ $xy - 2y = 1$ $xy = 1 + 2y$	M1
	$x = \dfrac{1+2y}{y}$	A1
	$f^{-1}(x) = \dfrac{1+2x}{x}$	A1
(b)	$gf(x) = \dfrac{1}{(x-2)^2}$	B1
(c)	$fg(x) = \dfrac{1}{x^2-2}$ $(x-2)^2 = x^2 - 2$ $x^2 - 4x + 4 = x^2 - 2$	M1
	$x = \dfrac{3}{2}$	A1
16	$AB = CD$ since opposite sides of a parallelogram are of equal length	B1
	Angles BAE and FCD are equal, since opposite angles in a parallelogram are equal	B1 B1
	By SAS rule, $\triangle BAE$ and $\triangle DCF$ are congruent	B1
	Therefore $BE = FD$ as required	

Question	Answer	Mark
17	Volume of hemisphere $= \dfrac{2}{3}\pi r^3 = \dfrac{2}{3}\pi\left(\sqrt{3}\right)^3 = \dfrac{2}{3}\pi 3\sqrt{3} = 2\sqrt{3}\pi$	M1 A1
	Volume of cone $= \dfrac{1}{3}\pi r^2 h = \dfrac{1}{3}\pi(3)2\sqrt{3} = 2\sqrt{3}\pi$	M1 A1
	Total volume $= 2\sqrt{3}\pi + 2\sqrt{3}\pi = 4\sqrt{3}\pi$	A1
18 (a)	$1.3^4 + 2 \times 1.3 - 7 = -1.54 < 0$ $1.5^4 + 2 \times 1.5 - 7 = 1.0625 > 0$	B1 B1
	There is a sign change, so there is a solution between $x = 1.3$ and $x = 1.5$	
(b)	$x_0 = 1.4$ $x_1 = 1.4316$	B1
	$x_2 = 1.4262$ $x_3 = 1.427$	B1 B1
19	$y = -\dfrac{b}{2a}$ $= -\left(\dfrac{-2}{-8}\right)$	M1
	$x = -\dfrac{1}{4}$	A1
	$y = 3 - 2\left(-\dfrac{1}{4}\right) - 4\left(-\dfrac{1}{4}\right)^2$	M1
	$= 3 + \dfrac{1}{2} - \dfrac{1}{4}$ $= \dfrac{13}{4}$	
	Maximum point is $\left(-\dfrac{1}{4}, \dfrac{13}{4}\right)$	A1

Set A – Paper 3

Question	Answer	Mark
1	$\tan 35° = \dfrac{12}{x}$	M1
	$x = \dfrac{12}{\tan 35°} = 17.1$ cm	A1
2	$\dfrac{£21120}{0.88} = £24\,000$	M1 A1
3	Using similar triangles	M1
	$\dfrac{x+12}{13} = \dfrac{12}{10.5}$	A1
	Solve to give $x = 2.86$ cm	A1

Question	Answer	Mark
4	Let $x = 0.1\ddot{2}\ddot{7} = 0.127272727...$	B1
	$10x = 1.27272727...$	B1
	$1000x = 127.272727...$	
	$990x = 126$	M1
	$x = \dfrac{126}{990}$	
	$= \dfrac{7}{55}$	A1
5 (a)	Equation – only valid for certain values of x	B1
(b)	Identity – true for all values of x	B1
(c)	Equation – only valid for certain values of x	B1
6	Distance = area under graph	M1
	$= \left(\dfrac{1}{2} \times 8 \times 5\right) + (22 \times 5) + \left(\dfrac{1}{2} \times 4 \times 5\right)$	A1
	$= 140\text{m}$	A1
7	Use $A = \dfrac{\theta}{360} \times \pi r^2$	M1
	$\theta = \dfrac{250 \times 360}{\pi \times 15^2}$	
	$= 127°$	A1
8 (a)	Summing areas under the bars	M1
	$25 + 14 + 16 + 4h = 70$	A1
	$4h = 15$	
	$h = 3.75$	A1
(b)	Sum of first two bars' area is 39, so median lies in 2nd bar	
	Suppose the median is x	
	Then by considering areas,	
	$25 + 7(x - 10) = 35$	M1
	Solve to give $x = 11.4$	A1
9	$12(x - 3) - 2(x - 2) = 3(x - 2)(x - 3)$	M1
	$12x - 36 - 2x + 4 = 3(x^2 - 5x + 6)$	
	$10x - 32 = 3x^2 - 15x + 18$	
	$3x^2 - 25x + 50 = 0$	A1
	$x = \dfrac{-b \pm \sqrt{b^2 - 4ac}}{2a}$	M1
	$x = \dfrac{25 \pm \sqrt{25^2 - 4 \times 3 \times 50}}{6}$	
	$x = 5$ or $x = 3.33$	A1 A1
10 (a)	$\mathbf{b} - \mathbf{a}$	B1
(b)	$\dfrac{1}{n+1}(\mathbf{b} - \mathbf{a})$	B1

Question	Answer	Mark
(c)	$\mathbf{a} + \dfrac{1}{n+1}(\mathbf{b} - \mathbf{a})$	M1
	$= \dfrac{(n+1)\mathbf{a} + (\mathbf{b} - \mathbf{a})}{n+1}$	
	$= \dfrac{n\mathbf{a} + \mathbf{b}}{n+1}$	A1
(d)	$\overrightarrow{OD} = \lambda \overrightarrow{OC} = \dfrac{\lambda n}{n+1}\mathbf{a} + \dfrac{\lambda}{n+1}\mathbf{b}$	B1
	Also $\overrightarrow{OD} = \overrightarrow{OA} + \overrightarrow{AD} = \mathbf{a} + \dfrac{2}{5}\overrightarrow{AF}$	
	$= \mathbf{a} + \dfrac{2}{5}\left(-\mathbf{a} + \dfrac{1}{2}\mathbf{b}\right) = \dfrac{3}{5}\mathbf{a} + \dfrac{1}{5}\mathbf{b}$	B1
	Equating coefficients and solving simultaneously	M1
	$\dfrac{\lambda n}{n+1} = \dfrac{3}{5}$ and $\dfrac{\lambda}{n+1} = \dfrac{1}{5}$	
	$\Rightarrow \lambda = \dfrac{3(n+1)}{5n}$ and $\lambda = \dfrac{n+1}{5}$	
	$\Rightarrow \dfrac{3(n+1)}{5n} = \dfrac{n+1}{5}$	
	$\Rightarrow n = 3$	A1
11 (a)	The graph shows a curve and not a straight line	B1
(b)	Draw a tangent line at the point on the curve where $x = 2$	M1
	Select two points on the line and calculate the gradient of the line using	
	$m = \dfrac{y_2 - y_1}{x_2 - x_1}$	M1
	Calculate $m \approx -0.4$	A1
	Deceleration ≈ 0.4 m/s²	A1
12 (a)	$x^4 - 5x^2 - 1 = 0$	
	$x^4 - 5x^2 + 3 = 4$	B1
	$x = -2.3$ or $x = 2.3$	B1 B1
(b)	$x^4 - 5x^2 - x + 2 = 0$	
	$x^4 - 5x^2 + 3 = x + 1$	B1
	$x = -2,\ x = -0.8,\ x = 0.6,\ x = 2.25$	B1 B1
13	RGB may be arranged in $3! = 6$ ways	B1
	Each arrangement has the probability occurring of	
	$\dfrac{2}{20} \times \dfrac{8}{19} \times \dfrac{10}{18} = \dfrac{4}{171}$	M1 A1
	Total probability is $6 \times \dfrac{4}{171} = \dfrac{24}{171}$	M1 A1

Question	Answer	Mark
14 (a)	$2x^2 - 6x + 1$	
	$= 2\left(x^2 - 3x + \dfrac{1}{2}\right)$	M1
	$= 2\left(\left(x - \dfrac{3}{2}\right)^2 - \dfrac{9}{4} + \dfrac{1}{2}\right)$	M1
	$= 2\left(\left(x - \dfrac{3}{2}\right)^2 - \dfrac{7}{4}\right)$	A1
	$= 2\left(x - \dfrac{3}{2}\right)^2 - \dfrac{7}{2}$	A1
(b)	Range is $f(x) \geqslant -\dfrac{7}{2}, f(x) \in \mathbb{R}$	B1 B1
15	$8 = A \times 5^{-20k}$ (1)	M1
	$1.6 = A \times 5^{-40k}$ (2)	M1
	Equation (1) divide equation (2)	A1
	$5 = 5^{20k}$	
	$20k = 1$	
	$k = \dfrac{1}{20}$	A1
	Substitute in (1)	
	$8 = A \times 5^{-1}$	
	$A = 40$	A1
16	Total surface area $= \pi r^2 + \pi r l$	M1
	$= \pi r^2 + 25\pi r = 600\pi$	A1
	$r^2 + 25r - 600 = 0$	
	$(r + 40)(r - 15) = 0$	M1
	$r = 15$	A1
	$h = \sqrt{25^2 - 15^2} = 20 \text{ cm}$	M1
	Volume of cone $V = \dfrac{1}{3}\pi r^2 h$	M1
	$V = \dfrac{1}{3} \times \pi \times 15^2 \times 20$	
	$= 1500\pi \text{ cm}^3$	A1

Question	Answer	Mark
17 (a)	$OA^2 = \left(2\sqrt{5}\right)^2 + 6^2 = 56$	M1
	Equation of circle is	
	$x^2 + y^2 = 56$	A1
(b)	Gradient of OA is $\dfrac{6}{2\sqrt{5}} = \dfrac{3}{\sqrt{5}}$	B1
	Gradient of AB is $-\dfrac{\sqrt{5}}{3}$	M1
	Equation of AB is $y = -\dfrac{\sqrt{5}}{3}x + c$	
	Substitute $x = 2\sqrt{5}$, $y = 6$	M1
	$6 = -\dfrac{\sqrt{5}}{3}\left(2\sqrt{5}\right) + c$	
	$c = \dfrac{28}{3}$	
	$y = -\dfrac{\sqrt{5}}{3}x + \dfrac{28}{3}$	A1
(c)	At B, $y = 0$, so $x = \dfrac{28}{\sqrt{5}} = \dfrac{28\sqrt{5}}{5}$	B1
	Area $OAB = \dfrac{1}{2}bh$	M1
	$= \dfrac{1}{2}\left(\dfrac{28\sqrt{5}}{5}\right)6$	
	$= \dfrac{84\sqrt{5}}{5}$	A1

Set B – Paper 1

Question	Answer	Mark	Comments
1	$x + 3$	B1	
2	2 and –3	B1	
3	$6^2 + 4^2$	M1	
	$6^2 + 4^2 = 52$, $\sqrt{52}$ cm	A1	
4	$6x - 12 + 8 = x$	M1	
	$5x = 4$	M1dep	
	$x = 0.8$ oe	A1	
5	Area of any face, i.e. 20×5 or 100 etc.	M1	
	$2 \times 100 + 2 \times 50 + 2 \times 200$	M1dep	
	700	A1	
6	$4x + 4 - 6x + 8$	M1	M1 for 3 terms correct
	$4x + 4 - 6x + 8$	A1	A1 for 4 terms correct
	$-2x + 12$	A1ft	ft on M1, e.g. $4x + 1 - 6x - 8 = -2x - 7$ is M1, A0, A1ft
7	$2x + 100 = 180$	M1	
	$360 \div 40$	M1dep	
	9	A1	
8 (a)	230 000	B1	
(b)	5×10^{-4}	B1	
(c)	1.6×10^8	B2	B1 for 16×10^7
9 (a)	–1.5 and 3	B2	B1 each answer
(b)	$(0.75, -6.1)$	B1	
10	$x + 2 = 2x - 1$	M1	
	$x = 3$	A1	
	$3 + 2$ or $2 \times 3 - 1$	M1dep	
	5	A1	
	25	A1	

Question	Answer	Mark	Comments
11	$x^2 + 2x + 1$ or $x^2 - 2x - 3$	M1	
	$x^3 - 3x^2 + 2x^2 - 6x + x - 3$	M1dep	
	$x^3 - x^2 - 5x - 3$	A1	
12	$\pi \times (r)^2 \times 6r$	M1	oe
	their $6\pi r^3 = 48\pi$	M1dep	
	2	A1	
13	$x \leqslant 6$	B1	
	$x + y \geqslant 7$	B1	
	$y \leqslant x + 1$	B1	
14	$27 + 9\sqrt{2} - 3\sqrt{8} - \sqrt{16}$	M1	oe
	$27 + 9\sqrt{2} - 6\sqrt{2} - 4$	A1	
	$23 + 3\sqrt{2}$	A1	
15	Vertical scale marked to at least 3.5 Bar between 5–10 to a height of 3 Bar between 10–20 to a height of 3.5 Bar between 20–35 to a height of 2 Bar between 35–45 to a height of 1.5 Bar between 45–50 to a height of 1	B3	B2 Scale marked and any two bars B1 Scale marked and any 1 bar
16 (a)	56°	B1	
(b)	60°	B1	
(c)	ACB stated or shown as 32	B1	
	CAB stated or shown as 90 (may be implied by working)	B1	
	58°	B1	
17	16	B2	B1 for $(\sqrt[3]{64})^2$ oe B1 for $\sqrt[3]{64} = 4$
18 (a)	24	B1	
(b)	31 and 17 seen	M1	
	14	A1	
(c)	Valid box plot with Median marked (ft their median) IQR marked (ft their IQR) Minimum value as 5 and maximum as 50	B2	B1 any 2 components

Question	Answer	Mark	Comments
19 (a)	$\mathbf{a} + \dfrac{3}{2}\mathbf{b}$	B1	
(b)	$\overrightarrow{BC} = \overrightarrow{BA} + \overrightarrow{AO} +$ $\overrightarrow{OC} = -\mathbf{a} + \dfrac{1}{2}\mathbf{b}$ or $-\dfrac{3}{2}\mathbf{b} - \mathbf{a} + \overrightarrow{OC}$ $= -\mathbf{a} + \dfrac{1}{2}\mathbf{b}$	M1	
	$2\mathbf{b}$	A1	
20	$x = 0.733333\ldots$ and $10x = 7.33333$	M1	
	$9x = 6.6$ or $\dfrac{66}{90}$	A1	
	$3\dfrac{11}{15}$	A1	
21	$\dfrac{x^2}{2} = 9$	M1	
	$x = 3\sqrt{2}$	A1	
	Hypotenuse $= 6$	A1	
	$6 + 2 \times 3\sqrt{2}$	M1	
	$6 + 6\sqrt{2}$	A1	
22	Tree diagram with at least 3 correct probabilities marked or P(R and B) + P(B and R)	M1	
	All correct probabilities identified as $\dfrac{7}{10}$, $\dfrac{3}{10}$, $\dfrac{6}{9}$ oe, $\dfrac{3}{9}$ oe, $\dfrac{7}{9}$ and $\dfrac{2}{9}$ or one of $\dfrac{7}{10} \times \dfrac{3}{9}$ or $\dfrac{3}{10} \times \dfrac{7}{9}$	A1	
	$\dfrac{7}{10} \times \dfrac{3}{9} + \dfrac{3}{10} \times \dfrac{7}{9}$	M1dep	
	$\dfrac{42}{90}$ or $\dfrac{7}{15}$	A1	

Question	Answer	Mark	Comments
23	$(2x + 3)(2x - 5)$ $(2x - 3)(x + 4)$ $2(x + 4)$ $(2x - 3)(2x + 3)$	B3	B2 three factorisations B1 two factorisations
	$\dfrac{2x - 5}{2}$	B1	
24	Gradient $AB = -\dfrac{1}{2}$	M1	
	Gradient perpendicular $= 2$	A1	
	Midpoint $AB = (5, 9)$	B1	
	$9 = 2 \times 5 + c$	M1	
	$y = 2x - 1$	A1	

Set B – Paper 2

Question	Answer	Mark	Comments
1	$(7, 6)$	B2	B1 either coordinate
2 (a)	alternate	B1	
(b)	$a + b = 180$	B1	
3	Correct translation i.e. $(1, 1) \rightarrow (-2, -3)$ etc.	B2	B1 for correct translation of one vector component
4	$6^2 + 11^2$	M1	
	$\sqrt{157}$	M1dep	
	$12.5\ldots$	A1	
5	$5 \times 145 + 9 \times 155 + 12 \times 165 + 8 \times 175 + 6 \times 185$ or 6610	M1	
	$6610 \div 40$	M1dep	
	165.25	A1	
6 (a)	Any product including a prime that makes 28	M1	
	$2 \times 2 \times 7$ or $2^2 \times 7$	A1	
(b)	$2 \times 2 \times 5 \times 7$	M1	
	140	A1	

Question	Answer	Mark	Comments
7	$4(x + 4) = 26$	M1	
	$4x = 10$	M1dep	
	2.5	A1	
8	0.85	B1	
	$238 \div 0.85$	M1	
	280	A1	
9	$36 \div 3$ or 12	M1	
	2×12 or 5×12	M1dep	
	24 and 60	A1	
10	$\sqrt{\dfrac{402}{\pi}}$ or 11.3…	M1	
	$11.3 \times \pi + 2 \times 11.3$	M1dep	
	[58, 58.2]	A1	
11	Arc from A cutting given line	M1	
	Arc centred on intersection and crossing original arc plus line drawn and angle 60° drawn	A1	
	60° angle bisected	A1	Angle must be between [26, 32]
12 (a)	$4x^2 - 8x + 3x - 6$	M1	
	$20x^2 - 25x - 30$	A1	
(b)	$2(x + a)(x + b)$	M1	$ab = \pm3$
	$2(x + 1)(x + 3)$	A1	oe e.g. $(2x + 2)(x + 3)$
13	Triangle between (3, 9), (4, 9) and (4, 7)	B3	B2 two vertices correct B1 rays marked through (5, 8)
14	30×1.6 or 48	M1	
	(their $48 - 40$) $\div 40$ ($\times 100$)	M1dep	
	20	A1	

Question	Answer	Mark	Comments
15 (a)	$(x + 3)^2$	M1	
	$(x + 3)^2 - 9$	M1dep	
	$(x + 3)^2 - 18$	A1	
(b)	$x + 3 = \sqrt{18}$	M1	
	$x = -3 \pm \sqrt{18}$	A1	
16	$2(4x - 1) - 3(x + 1)$	M1	
	$5x - 5 =$	A1	
	$(4x - 1)(x + 1)$ or $4x^2 + 4x - x - 1$	M1	
	$4x^2 - 2x + 4$	A1	
17 (a)	$y = kx^2$ and $20 = k \times 2^2$	M1	
	$k = 5$	A1	
	500	A1	
(b)	$5 = 5 \times x^2$	M1	
	±1	A1	Condone omission of $\pm$
18	$x(x - 6) + 2x + x - 4 + x + 30 = 146$	M1	
	$x^2 - 2x - 120 = 0$	A1	
	$(x - 12)(x + 10) = 0$	A1	
	$x = 12$	A1	
	$\dfrac{8}{146}$ or $\dfrac{4}{73}$	A1	
19 (a)	$\cos x = \dfrac{10^2 + 7^2 - 13^2}{2 \times 10 \times 7}$	M1	
	$-\dfrac{1}{7}$	A1	
	98.2	A1	
(b)	$\dfrac{1}{2} \times 7 \times 10 \times \sin$ (their 98.2)	M1	
	34.6…	A1	

Question	Answer	Mark	Comments
20	$\dfrac{6(3+\sqrt{3})}{(3-\sqrt{3})(3+\sqrt{3})}$	M1	
	$\dfrac{18+6\sqrt{3}}{9-3}$	A1	
	$\dfrac{6(3+\sqrt{3})}{6}$	A1	
21	11.5 or 12.5 or 8.25 or 8.35	M1	
	11.5 and 12.5 and 8.25 and 8.35	M1dep	
	$8.25 \div 12.5$ or $8.35 \div 11.5$	M1	
	Upper 46.6	A1	
	Lower 41.3	A1	
22 (a)	Tangent drawn at 3	M1	
	y-step and x-step measured	M1dep	
	[0.7, 1.1]	A1ft	ft their tangent
(b)	Attempt to calculate area under curve	M1	
	[75, 85]	A1ft	ft their area
	Their area $\div$ 10	M1dep	
	[7.5, 8.5]	A1	

Set B – Paper 3

Question	Answer	Mark	Comments
1	729	B1	
2	125	B1	
3 (a)	x^9	B1	
(b)	x^{10}	B1	
4	$\begin{pmatrix}10\\4\end{pmatrix}$	B2	B1 for each component
5	$-2, -1, 0, 1, 2, 3$	B2	B1 for -3, $-2, -1, 0, 1, 2, 3$ B1 for -2, $-1, 0, 1, 2, 3, 4$

Question	Answer	Mark	Comments
6	$1^2 + 2^2$	M1	
	$\dfrac{\sqrt{3}}{2}$	A1	
7		B2	B1 for any enlargement that reduces the size of the shape and keeps the side in relative ratio. B1 for any 3 sides correct.
8	$1.5 \div 2$	M1	
	0.75	A1	
9 (a)	$\dfrac{4}{10}$ marked on red and $\dfrac{6}{10}$ marked on blue	B1	
(b)	$\dfrac{4}{10} \times \dfrac{4}{10}$ or $\dfrac{6}{10} \times \dfrac{6}{10}$	M1	
	$\dfrac{4}{10} \times \dfrac{4}{10} +$ $\dfrac{6}{10} \times \dfrac{6}{10}$	M1dep	
	0.52	A1	oe
10	$3x + 2y = 2$ and $3x + 12y = 27$ or $6x + 4y = 4$ and $x + 4y = 9$	M1	
	$x = -1$	A1	
	$y = 2.5$	A1	
11 (a)	$(x + 5)(x - 5)$	B1	
(b)	$x^2 + 4x + 4$ or $x^2 + 2x + 1$	M1	$(x + 2 + x + 1)$ $(x + 2 - (x + 1))$
	$x^2 + 4x + 4 -$ $(x^2 + 2x + 1)$	M1 dep	$(2x + 3)(1)$
	Shows subtraction of terms clearly	A1	

Question	Answer	Mark	Comments
12 (a)	$12 \times \sin 32 = 6.359...$	B1	
(b)	$\pi \times 6.36 \times 12$	M1	
	[236.6, 240]	A1	
13	0.92	B1	
	0.92^n for $n > 3$	M1	
	9 years	A1	Accept just over 8 or between 8 and 9
14	A C B	B2	B1 for 1 correct
15	$4x^4y^6$	B2	B1 for 2 parts correct
16 (a)	B and D	B1	
(b)	A and D	B1	
17 (a)	30 and 38	B2	B1 each
(b)	Works out second difference as 2	M1	
	Subtracts n^2 from series, i.e. 5, 6, 7, 8, 9 …	M1dep	
	Identifies $n + 4$ as linear sequence	A1	
	$n^2 + n + 4$	A1	
18	Shows reflected triangle B at (7, 2), (7, 4) and (9, 4)	M1	
	Shows reflected triangle C at (1, 4), (3, 4) and (3, 2)	M1dep	
	Rotation, 180°, about (5, 6)	A2	A1 for 2 parts. Accept reflection in line $y + x = 11$ oe
19	$\dfrac{x}{\sin 78} = \dfrac{11}{\sin 65}$	M1	
	$x = \dfrac{11 \times \sin 78}{\sin 65}$	M1dep	
	[11.87, 11.9]	A1	
20	$14 \div 8$ or 1.75	M1	
	$540 \times$ (their 1.75)^3	M1dep	
	2890	A1	

Question	Answer	Mark	Comments
21	$AC = \sqrt{8^2 + 6^2}$ or 10	M1	
	$CX = 5$	A1	
	$VX = \sqrt{12^2 - 5^2}$ or $\sqrt{119}$ or 10.9...	M1dep	
	Angle $VCX =$ $\sin^{-1}(10.9 \div 12)$	M1dep	Can use cos or tan
	[65, 65.4]	A1	
22 (a)	$b = \sqrt[3]{2a - 3}$	B1	
(b)	−1	B1	
(c)	−1.89	B2	B1 for any further iterations or 1.89...
23 (a)	$x^2 + y^2 = 16$	B1	
(b)	Angle $= \tan^{-1}(2)$ or 63.43...	M1	
	(their 63.43 $\div$ 360) $\times$ 2 $\times \pi \times$ their radius	M1dep	
	[4.36, 4.43]	A1	
24	$\dfrac{4}{9}x$	M1	
	$\dfrac{4}{9}x + 7$	M1dep	
	$\dfrac{4}{9}x + 7 = \dfrac{x+7}{2}$	M1dep	
	$\dfrac{1}{18}x = \dfrac{7}{2}$	M1dep	
	63	A1	T&I B1 for correct answer
25 (a)	$\dfrac{x+1}{3}$	B2	B1 for numerator of $3(x + 1)$ B1 for $\dfrac{x-1}{3}$
(b)	$3(x^2 + 2) - 1$	M1	
	$3x^2 + 5$	A1	
26	$x^2 + (x + 3)^2$	M1	
	$x^2 + x^2 + 6x + 9 = x + 12$	A1	
	$2x^2 + 5x - 3 = 0$	M1	
	$(2x - 1)(x + 3)$	A1	
	$\left(\dfrac{1}{2}, 3\dfrac{1}{2}\right)$ and (−3, 0)	A1	

BLANK PAGE

BLANK PAGE